The Cornfield Maze Mystery

Companion Guide

Educators and Parents Resource

Teaching Materials & Resources

Supporting inquiry-based learning about food systems, agriculture, and health for ages 8-12

By David Upthegrove

Published by Upthegrove Press

ISBN 9798993490441

Cross-reference: The Cornfield Maze Mystery (Story Book)

Teacher Quick Reference

Quick unit snapshot:

- • Ages: 8-12
- • Timeframe: 2-4 class sessions (45-60 min each)
- • Core activities: read-aloud, hands-on experiments, research projects, poster presentations
- • Key aims: inquiry-based learning, media literacy, food-system awareness, critical thinking; avoid alarm, emphasize evidence-informed choices
- Download free PDF resource available at: www.thecornfieldmazemystery.com

Suggested teacher script: "Science is a process. Some studies show possible effects; others show different results. We will look at age-appropriate evidence and focus on practical steps to make informed choices while research continues."

Table of Contents

Section 1: Quick Start Guide 1

Section 2: Teacher-Facing Lesson Plans (Ages 8-12) 2

- Overview & Learning Objectives 2

- Materials & Timeframe 4

- Lesson Procedures Session-by-Session 5

- Assessment & Rubrics 11

- Differentiation Strategies 12

- Student Worksheets (Ready to Print) 13

- Answer Key 24

Section 3: Background Information for Educators 25

- GMO Corn and Herbicide Tolerance 25

- Glyphosate and Other Herbicides 27

- Health Concerns and Scientific Debate 28

- Ultra-Processed Foods and Health 29

- Gut Microbiome and Diet 31

- Environmental Impacts 33

Section 4: Parent Communication Materials 35

- Sample Parent Letter 35

- FAQ for Parents 37

- At-Home Extension Activities 39

Section 5: Additional Resources 43

- Expanded Glossary with Pronunciations 43

- Bibliography and Further Reading 53

- Community Connections 54

- Extension Projects 56

- Conclusion 58

Section 1: Quick Start Guide

Welcome, Educators!

This companion guide supports the children's book The Cornfield Maze Mystery providing ready-to-use lesson plans, student activities, and background information. The materials are designed to:

- ✓ Spark curiosity about food systems without causing alarm

- ✓ Support inquiry-based, age-appropriate learning

- ✓ Align with Common Core and NGSS standards

- ✓ Provide balanced, science-based information

- ✓ Engage families in meaningful conversations about food

How to Use This Guide

For a 2-4 class session unit: Go straight to Section 2 (Teacher Lesson Plans) for complete, ready-to-implement lessons.

For parent engagement: Use Section 4 (Parent Communication Materials) for newsletters, parent nights, or take-home activities.

For background & deeper understanding: Read Section 3 (Background Information) to understand the science and current debates around agricultural chemicals, GMOs, and food processing.

Safety & Sensitivity Note

This curriculum involves NO chemical handling. All experiments are observational and use safe, common materials. The book and lessons present agricultural chemicals (specifically herbicides like glyphosate) and ultra-processed foods as real-world examples, focusing on empowerment through knowledge and practical actions rather than fear.

What You'll Need
- One class set of the children's book (or single copy for read-aloud)
- Student worksheets from Section 2 (print-ready)
- Basic supplies for experiments (seeds, soil, jars, water, food labels)
- Optional: materials for composting demonstration
- Poster materials for final project

Section 2: Teacher-Facing Lesson Plans

"From Seed to Spoon: Understanding Our Food System" — Ages 8-12

Overview

This plan supports a comprehensive unit (2-4 class sessions) built around the narrative about three children exploring corn production, farming practices, food processing, and health. Lessons are inquiry-driven, hands-on, and designed to spark curiosity without alarm. Activities include read-aloud with guided discussion, multiple classroom experiments, research projects, and presentations that can be completed in class or as homework.

Note: Download Free PDF worksheets and additional materials at:

www.thecornfieldmazemystery.com

Learning Objectives

Students will explain, in their own words, the journey of corn from field to table and identify the main steps in food production and processing.

Students will identify ways agricultural chemicals and food processing affect the food system and discuss evidence-based actions to make informed food choices.

Students will conduct hands-on experiments, observe and record changes, and use evidence to draw conclusions.

Students will complete a research project that communicates practical actions to support healthier food choices and environmental sustainability.

Students will practice critical thinking, source evaluation, and respectful dialogue about complex scientific and social issues.

Standards Alignment

NGSS (Next Generation Science Standards):

3-LS4-3: Construct an argument with evidence that in a particular habitat some organisms can survive well, some survive less well, and some cannot survive at all. (Farm ecosystem observations)

3-5-ETS1-1/2: Define a simple design problem and generate/test solutions. (Designing experiments, data collection)

5-LS2-1: Develop a model to describe movement of matter among plants, animals, decomposers, and the environment. (Food system, soil health, gut microbiome)

MS-LS2-1: Analyze and interpret data to provide evidence for the effects of resource availability on organisms and populations. (Agricultural practices and biodiversity)

Common Core ELA (Grades 3-6):

CCSS.ELA-LITERACY.RI.4.3: Explain events, procedures, ideas, or concepts in a historical, scientific, or technical text. (Tracing seed-to-spoon steps)

CCSS.ELA-LITERACY.W.4.7: Conduct short research projects that build knowledge through investigation of different aspects of a topic. (Science fair project)

CCSS.ELA-LITERACY.SL.4.1: Engage effectively in a range of collaborative discussions. (Interviews, team presentations, respectful dialogue)

CCSS.ELA-LITERACY.W.5.1: Write opinion pieces on topics, supporting a point of view with reasons and information. (Persuasive writing about food choices)

Health/Safety Standards (adapt to local requirements):

Promote safe handling of household and garden materials with adult supervision

Encourage balanced diets and consulting healthcare professionals for medical concerns

Develop media literacy and critical evaluation of health claims

Materials Needed

One class set of The Cornfield Maze and the Seeds of Change (or single copy for read-aloud)

Student worksheets (see pages 12-18) — print-ready

Chart paper / whiteboard and markers

Materials for Soil & Plant Experiments: soil samples, clear jars, bean seeds, small pots, water, labels, magnifying glasses, rulers

Materials for Food Label Activity: empty food packages or printed labels, highlighters, data collection sheets

Materials for Composting Demo: small bin, food scraps (vegetable peels), shredded paper, soil

Materials for Microbial Activity Jars: clear jars, vegetable scraps, water, labels

Poster paper or slide template, markers, colored pencils

Internet/library access for research (optional)

Optional: photos showing stages (seed, field, combine, processing plant, grocery store, meal)

Timeframe

Session 1 (45-60 min): Read-aloud + guided discussion + vocabulary + comprehension worksheet

Session 2 (45-60 min): Soil Detectives experiment + observations + data recording

Session 3 (45-60 min): Food Label Detective activity + ultra-processed foods discussion

Session 4 (30-45 min): Plant Growth & Composting experiment setup (ongoing observations over 2-4 weeks)

Session 5 (45-90 min): Research project planning + poster creation

Session 6 (45-60 min): Presentations + reflection + action planning

Note: Sessions can be adapted, combined, or spread over multiple weeks depending on your schedule and student needs.

Lesson Procedures Session-by-Session

Session 1 — Read & Discuss (45-60 min)

1. Warm-up (5 min): Ask students to name foods that contain corn or corn-derived ingredients. Record responses on chart paper. Most students will be surprised by how many foods contain corn!

2. Read-aloud (20-25 min): Read the narrative aloud (or have students follow along). Pause at key moments (planting, farm tour, processing plant visit, grocery store expedition, science fair) to check comprehension and invite predictions.

3. Guided discussion (15-20 min): Use discussion questions:

• What surprised you most about the story?

• How did Mary, Ana, and Ben investigate their questions? What methods did they use?

• What did you learn about where corn comes from and how it's processed?

• What are GMOs? Why do some farmers use them?

• What are ultra-processed foods? Why might they affect health?

• What choices do people have at different levels (farmers, companies, families, kids)?

• What questions do you still have?

4. Vocabulary introduction (10 min): Review key terms from the glossary (see Section 5). Provide simple definitions and pronunciations. Have students practice using terms in sentences.

5. Worksheet time (10 min): Hand out and complete the comprehension worksheet (Part A) in pairs or individually.

Session 2 — Soil Detectives Experiment (45-60 min)

This hands-on activity mirrors the Corn Crew's soil investigation in the story.

1. Introduction (5 min): Review the part of the story where the kids compare soil from different areas. Explain that students will become soil detectives.

2. Safety talk (3 min): Explain safe soil handling. Optional gloves. Wash hands after handling soil.

3. Soil collection (10 min): If possible, take students outside to collect soil samples from 2-3 different locations (e.g., garden bed, under trees, near pavement). If not possible, provide pre-collected samples. Students label jars clearly.

4. Observation & testing (20 min): Students observe and record:

• Color (dark brown, light tan, gray, etc.)

• Texture (sandy, clay-like, crumbly)

• Smell (earthy, no smell, other)

• Visible organisms (worms, insects, roots)

• Moisture level

• Optional: Simple pH test using vinegar (fizzing = alkaline) or baking soda solution (fizzing = acidic)

5. Data recording (10 min): Students complete observation worksheet with drawings and descriptions.

6. Class discussion (10 min): What differences did you notice? What might cause these differences? Why does healthy soil matter for growing food? How do farming practices affect soil?

This activity mirrors Ana's pantry investigation in the story.

1. Introduction (5 min): Review the part of the story where Ana investigates food labels and discovers corn ingredients everywhere. Explain that students will do the same investigation.

2. Label reading lesson (10 min): Teach students how to read ingredient lists. Explain that ingredients are listed in order by weight. Introduce common corn-derived ingredients:

• Corn syrup, high-fructose corn syrup (HFCS)

• Corn starch, modified corn starch

• Corn oil, vegetable oil (often includes corn)

• Corn meal, corn flour

• Dextrose, maltodextrin, glucose syrup

• Citric acid (often corn-derived)

• Xanthan gum

3. Investigation activity (20 min): In small groups, students examine 5-10 food packages (or printed labels). They complete a data sheet recording:

• Product name

• Number of ingredients

• Corn-derived ingredients found

• Added sugars (yes/no, how much)

• Classification: whole food, processed, or ultra-processed

4. Data analysis (10 min): As a class, compile results. What percentage of products contained corn ingredients? Which categories had the most? Discuss why corn is so common in processed foods (cheap, versatile, subsidized).

5. Ultra-processed foods discussion (10 min): Define ultra-processed foods. Discuss health concerns based on research (obesity, diabetes, heart disease, digestive issues). Emphasize that occasional consumption is fine, but a diet heavy in ultra-processed foods may affect health.

Session 4 — Plant Growth & Composting Experiment (30-45 min setup)

This experiment mirrors the Corn Crew's plant growth comparison in the story.

1. Introduction (5 min): Review the part of the story where the kids test plant growth in different soil conditions. Explain that students will conduct a similar experiment.

2. Experimental design (10 min): Discuss the scientific method. Students will plant bean seeds in three conditions:

• Pot 1 (Control): Regular soil, water, sunlight

• Pot 2 (Compost): Soil with added compost, water, sunlight

• Pot 3 (Stress): Soil with slightly less water or light (DO NOT use chemicals)

Discuss variables: What stays the same? What changes? How will we measure results?

3. Planting (15 min): Students work in groups to plant seeds, label pots clearly, and place in designated growing area.

4. Observation schedule (5 min): Create a class schedule for watering and measuring. Students will measure height, count leaves, and photograph plants every 3-4 days for 2-4 weeks.

5. Composting demonstration (optional, 10 min): Set up a small classroom compost bin. Explain what can be composted (vegetable scraps, fruit peels, shredded paper, leaves) and what cannot (meat, dairy, oils). Discuss how composting reduces waste and creates healthy soil.

Note: This experiment runs in the background over 2-4 weeks. Students record observations regularly and analyze results at the end.

Session 5 — Research Project: Food Choices Poster (45-90 min)

This project mirrors the Corn Crew's science fair project.

1. Project introduction (10 min): Explain the project goals and rubric. Students will create a poster or presentation answering the question: 'How can we make informed food choices that support our health and the environment?'

2. Topic selection (10 min): Students choose a focus area:

• Organic vs. conventional farming

• Reducing ultra-processed foods in your diet

• Supporting local farmers and food systems

• Composting and reducing food waste

• Reading labels and understanding ingredients

• The environmental impact of different farming practices

• Gut health and the microbiome

3. Research time (20-30 min): Students work in pairs or small groups to research using library books, approved websites, or provided materials. They take notes and identify key facts, statistics, and action steps.

4. Poster creation (30-40 min): Students design and create posters including:

• Title and names

• Key facts and evidence

• Visuals (drawings, charts, photos)

• Practical action steps

• Sources cited

5. Peer review (optional, 10 min): Students exchange posters with another group for feedback before final revisions.

Session 6 — Presentations & Reflection (45-60 min)

1. Presentations (30-40 min): Each group presents their poster (3-5 minutes). Audience asks questions and provides positive feedback.

2. Plant experiment results (10 min): Review data from the plant growth experiment. Which pot grew best? Why? What does this tell us about soil health and growing conditions?

3. Reflection (10 min): Students complete individual reflection worksheets:

• What did you learn about the food system?

• What surprised you most?

• What action will you take based on what you learned?

• What questions do you still have?

4. Action planning (5 min): As a class, brainstorm one collective action (e.g., start a school garden, organize a food label awareness campaign, invite a local farmer to speak).

5. Celebration (5 min): Celebrate student learning! Display posters in the classroom or hallway for others to see.

Assessment & Rubrics

Formative Assessment

✓ Participation in read-aloud and guided discussions

✓ Completion of worksheets (comprehension, observation, data collection)

✓ Experiment observations and hypothesis formation

✓ Engagement in group activities and peer collaboration

Summative Assessment: Research Poster/Presentation

Scored on 16-point scale (4 categories × 4 points each):

Category	Excellent (4)	Proficient (3)	Developing (2)	Beginning (1)
Content Accuracy	All facts accurate, well-researched, sources cited	Most facts accurate, sources cited	Some inaccuracies, few sources	Many inaccuracies, no sources
Visual Design	Clear, creative, well-organized, engaging	Clear and organized	Somewhat unclear or disorganized	Unclear, hard to follow
Action Steps	Specific, practical, evidence-based actions	Clear actions with some detail	Vague or impractical actions	No clear actions
Presentation	Clear, confident, answers questions well	Clear, answers most questions	Unclear at times, struggles with questions	Unclear, cannot answer questions

Total possible: 16 points

13-16 = A (Excellent)

10-12 = B (Proficient)

7-9 = C (Developing)

4-6 = D (Beginning)

Differentiation Strategies

For Emerging Readers/Writers:
- Provide sentence starters and word banks for written responses
- Use visual supports (diagrams, photos, graphic organizers)
- Allow audio recording of observations and reflections
- Pair with stronger readers for partner reading
- Offer simplified vocabulary lists with pictures

For Advanced Learners:
- Encourage deeper research using primary sources (scientific studies, government reports)
- Challenge students to design their own experiments with multiple variables
- Invite students to write persuasive essays or create public service announcements
- Provide extension activities (interview local farmers, analyze food policy, calculate environmental footprints)
- Encourage leadership roles in group projects

For English Language Learners:
- Pre-teach key vocabulary with visual aids and translations
- Provide bilingual glossaries when possible
- Use hands-on activities to support language development
- Allow responses in native language with translation support
- Pair with bilingual peers for support

For Diverse Learning Needs:
- Offer choice in how students demonstrate learning (poster, presentation, video, model, written report)
- Provide flexible seating and movement breaks during longer sessions
- Use multi-sensory approaches (visual, auditory, kinesthetic)
- Break complex tasks into smaller, manageable steps with checklists
- Provide extra time for completion when needed

Student Worksheets (Ready to Print)

Note: The following worksheets are designed to be printed and distributed to students. They align with the lesson plans in Section 2.

Worksheet 1: Story Comprehension

Name: _____   Date: _____

After reading The Cornfield Maze Mystery, answer the following questions:

1. Who are the three main characters in the story? What are their names and ages?

_____

_____

2. What did the Corn Crew decide to investigate for their science fair project?

_____

_____

3. List three places the children visited during their investigation:

a) _____

b) _____

c) _____

4. What is a GMO? Why do some farmers use GMO corn?

_____

_____

_____

5. What are ultra-processed foods? Give three examples from the story.

_____

_____

_____

6. What pattern did Ana notice about her mother's stomach problems?

_____

_____

7. What advice did Dr. Lopez give about making healthy food choices?

_____

_____

_____

8. What did the Corn Crew learn about asking questions and doing research?

_____

_____

_____

Name: _____   Date: _____

Collect soil samples from two different locations. Observe carefully and record your findings:

Observation	Sample 1 Location: _____	Sample 2 Location: _____
Color		
Texture (sandy, clay, crumbly)		
Smell		
Moisture (dry, damp, wet)		
Visible organisms (worms, insects)		
Other observations		
Drawing (sketch what you see)		

Analysis Questions:

1. What differences did you notice between the two soil samples?

_____

_____

2. Which soil sample do you think would be better for growing plants? Why?

_____

_____

3. What might cause the differences you observed?

_____

_____

Name: _____ Date: _____

Examine food labels and record your findings:

Product Name	# of Ingredients	Corn Ingredients Found	Added Sugars?	Whole/Processed/Ultra-Processed

Common corn-derived ingredients to look for:

• Corn syrup, high-fructose corn syrup (HFCS)

• Corn starch, modified corn starch

• Corn oil, vegetable oil

• Dextrose, maltodextrin, glucose syrup

• Corn meal, corn flour

Analysis Questions:

1. How many of the products you examined contained corn-derived ingredients?

_____

2. Were you surprised by how common corn ingredients are? Why or why not?

_____

_____

3. Which product had the most ingredients? Was it whole, processed, or ultra-processed?

_____

_____

4. Based on what you learned, what is one food swap you could make to eat healthier?

_____

_____

Name: _____   Date Started: _____

Track your plant growth over 2-4 weeks. Measure and record every 3-4 days:

Date	Pot 1 (Control) Height (cm)	Pot 2 (Compost) Height (cm)	Pot 3 (Stress) Height (cm)

Additional Observations:

Leaf color: _____

Number of leaves: _____

Insects or other visitors: _____

Other notes: _____

_____

Final Analysis:

1. Which pot grew the tallest? Why do you think this happened?

_____

_____

2. What does this experiment teach us about soil health and plant growth?

_____

_____

3. How might this relate to farming practices and food production?

_____

_____

Name(s): _____ Date: _____

Use this worksheet to plan your research poster or presentation:

Topic: _____

Research Question: What do you want to find out?

_____

_____

Key Facts (List at least 3 important facts you discovered):

1. _____

_____

2. _____

_____

3. _____

_____

Sources (Where did you find your information?):

1. _____

2. _____

3. _____

Action Steps (What can people do based on your research?):

1. _____

2. _____

3. _____

Visual Ideas (What images, charts, or drawings will you include?):

_____

_____

Name: _____ Date: _____

Reflect on what you learned during this unit:

1. What did you learn about the food system (how food gets from farms to your table)?

_____

_____

_____

2. What surprised you most during this unit?

_____

_____

_____

3. What is one thing you will do differently based on what you learned?

_____

_____

_____

4. What questions do you still have about food, farming, or health?

_____

_____

_____

5. How can you share what you learned with your family or community?

_____

_____

_____

6. On a scale of 1-5, how much did you enjoy this unit? (Circle one)

1 (not at all)   2 (a little)   3 (somewhat)   4 (quite a bit)   5 (very much)

7. What did you like best about this unit?

_____

_____

Answer Key for Worksheets

1. The three main characters are Mary (11), Ana (11), and Ben (10). They call themselves the Corn Crew.

2. They decided to investigate corn production from seed to table, including GMOs, herbicides, food processing, and health effects.

3. Three places they visited: Uncle Ray's farm, Heartland Processing Plant, Miller's Grocery Store (also acceptable: Dr. Lopez's office, school library)

4. GMO stands for Genetically Modified Organism. It's a plant whose DNA has been changed by scientists. Some farmers use GMO corn because it can tolerate herbicides or resist insect pests, which helps them control weeds and reduce crop damage.

5. Ultra-processed foods are made mostly from industrially processed substances like oils, fats, sugars, and starches, with added flavors, colors, and preservatives. Examples from the story: soda, chips, instant noodles, packaged snacks, many breakfast cereals, candy.

6. Ana noticed that her mother had more stomach problems on days when she ate a lot of processed foods, and felt better on days when she ate simpler, whole foods.

7. Dr. Lopez advised eating more whole foods (fruits, vegetables, whole grains, beans), reading labels, cooking more at home, and not aiming for perfection—an 80/20 approach is realistic.

8. The Corn Crew learned that asking questions is how we learn, that science doesn't always have simple answers, that it's important to look at evidence from multiple sources, and that everyone has power to make informed choices.

Worksheets 2-6: Observation-Based

Worksheets 2-6 are based on student observations and personal responses. There are no single correct answers. Assess based on:

• Completeness of observations and data recording

• Evidence of careful observation and critical thinking

• Ability to draw reasonable conclusions from evidence

• Thoughtful reflection and personal connection to the material

Section 3: Background Information for Educators

This section provides educators with essential background knowledge about the topics covered in The Cornfield Maze Mystery. The information is presented in an accessible, balanced manner to support informed classroom discussions.

GMO Corn and Herbicide Tolerance

Genetically Modified Organisms (GMOs) are plants, animals, or microorganisms whose DNA has been altered using genetic engineering techniques. In the United States, approximately 90% of corn grown is genetically modified.

Two Main Types of GMO Corn:

1. Herbicide-Tolerant Corn: Engineered to survive exposure to specific herbicides, most commonly glyphosate (brand name Roundup). Farmers can spray entire fields to kill weeds while the corn survives. This trait is created by inserting genes that produce enzymes resistant to the herbicide's effects.

2. Bt Corn: Engineered to produce a protein from the bacterium Bacillus thuringiensis (Bt) that is toxic to certain insect pests, particularly corn borers and rootworms. This reduces the need for insecticide spraying.

Why Farmers Use GMO Corn:

• Weed control: Herbicide-tolerant crops simplify weed management

• Reduced crop loss: Bt corn protects against insect damage

• Labor savings: Less need for mechanical cultivation or hand weeding

• Reduced insecticide use: Bt corn can reduce the need for spraying

• Economic pressure: Competitive farming economics favor high-yield, low-labor methods

Concerns about GMO Corn:

• Increased herbicide use: Some studies show herbicide use has increased as weeds develop resistance

• Environmental impacts: Effects on non-target insects, soil microbes, and water quality

• Corporate control: Most GMO seeds are patented by large corporations

• Lack of long-term studies: Questions remain about long-term environmental and health effects

• Labeling and transparency: In the U.S., GMO labeling is not required for most foods

Balanced Perspective: Major scientific organizations (National Academy of Sciences, American Medical Association, World Health Organization) have concluded that currently approved GMO crops are safe to eat. However, concerns about environmental impacts, corporate control of seeds, and the need for ongoing research remain valid topics for discussion.

Glyphosate and Other Herbicides

Herbicides are chemicals used to control unwanted plants (weeds). Glyphosate is the most widely used herbicide in the world, applied to crops, lawns, gardens, and public spaces.

How Glyphosate Works:

Glyphosate blocks an enzyme called EPSPS, which plants need to produce certain amino acids. Without these amino acids, plants cannot grow and eventually die. Humans and animals don't have this enzyme pathway, which is why glyphosate was initially considered safe for non-plant organisms.

Glyphosate in the Food System:

• Applied to GMO crops during the growing season

• Sometimes used as a pre-harvest desiccant (drying agent) on wheat, oats, and other crops

• Can leave residues on food, though usually below regulatory limits

• Water-soluble, so it can wash into groundwater and surface water

• Breaks down in soil over days to weeks, depending on conditions

Health and Safety Debates:

The safety of glyphosate is one of the most contentious scientific debates today:

Regulatory Agencies (EPA, FDA, EFSA): These agencies have concluded that glyphosate is unlikely to be carcinogenic to humans at typical exposure levels and that residues in food are within safe limits.

International Agency for Research on Cancer (IARC): In 2015, IARC classified glyphosate as "probably carcinogenic to humans" (Group 2A) based on limited evidence in humans and sufficient evidence in animals.

Independent Research: Some studies suggest glyphosate may:

• Disrupt gut bacteria (since many gut microbes have the EPSPS enzyme)

• Act as an endocrine disruptor at low doses

• Contribute to health problems when combined with other chemicals (synergistic effects)

• Accumulate in the body with repeated exposure

Litigation: Thousands of lawsuits have been filed by people claiming glyphosate exposure caused their cancer (non-Hodgkin lymphoma). Some juries have awarded large damages, though these cases are being appealed.

Other Common Herbicides:

• Atrazine: Used on corn; banned in Europe due to water contamination concerns; linked to endocrine disruption

• 2,4-D: Component of Agent Orange; still used on lawns and crops; concerns about drift and health effects

• Dicamba: Prone to drift damage; has damaged millions of acres of non-target crops

Teacher Note: When discussing herbicides with students, emphasize that scientists are still researching these questions, that different studies sometimes reach different conclusions, and that making informed choices means considering multiple sources of evidence.

Ultra-Processed Foods and Health

Ultra-processed foods (UPFs) are industrial formulations made mostly from substances extracted from foods (oils, fats, sugar, starch, protein) plus additives like flavors, colors, emulsifiers, and preservatives. They typically contain little to no whole foods.

Examples of Ultra-Processed Foods:

• Sodas and sweetened drinks

• Packaged snacks (chips, crackers, cookies, candy)

• Instant noodles and soups

• Many breakfast cereals

• Packaged breads and baked goods with long ingredient lists

• Frozen meals and pizza

• Processed meats (hot dogs, chicken nuggets)

• Sweetened yogurts with many additives

Why Ultra-Processed Foods Are Concerning:

Large-scale studies from multiple countries have found strong associations between high UPF consumption and:

• Obesity and weight gain

• Type 2 diabetes

• Cardiovascular disease

• Some cancers

• Digestive problems and inflammatory bowel disease

• Depression and anxiety

• Earlier death from all causes

Why UPFs May Harm Health:

1. Nutrient Poor: Low in fiber, vitamins, minerals, and beneficial plant compounds

2. Hyperpalatable: Engineered to be extremely tasty, leading to overconsumption

3. Rapidly Digested: Cause blood sugar spikes and crashes, leading to hunger and overeating

4. Additives: May affect gut bacteria, metabolism, and inflammation

5. Displacement: Replace whole foods that would provide health benefits

6. Marketing: Heavily marketed to children and low-income communities

The Role of Corn in Ultra-Processed Foods:

Corn-derived ingredients are ubiquitous in ultra-processed foods:

• High-fructose corn syrup (HFCS): Cheap sweetener in sodas, baked goods, condiments

• Corn starch: Thickener and filler

• Corn oil: Inexpensive cooking oil

• Maltodextrin, dextrose: Sweeteners and bulking agents

• Modified corn starch: Texture modifier

U.S. agricultural policy subsidizes corn production, making corn-derived ingredients very cheap. This economic reality drives their widespread use in processed foods.

Practical Guidance for Families: Focus on eating more whole foods (fruits, vegetables, whole grains, beans, nuts, minimally processed meats and dairy) rather than obsessing over avoiding all processed foods. An 80/20 approach—mostly whole foods with some processed foods—is realistic and sustainable.

Gut Microbiome and Diet

The gut microbiome refers to the trillions of bacteria, fungi, viruses, and other microorganisms living in our digestive system. These microbes play crucial roles in digestion, immune function, mental health, and overall well-being.

Why the Gut Microbiome Matters:

- Digests fiber and produces beneficial short-chain fatty acids

- Produces vitamins (B vitamins, vitamin K)

- Trains and regulates the immune system

- Protects against harmful bacteria

- Communicates with the brain via the gut-brain axis

- Influences mood, behavior, and mental health

What Affects the Gut Microbiome:

Diet: The most important factor. Fiber-rich whole foods feed beneficial bacteria. Ultra-processed foods, low fiber, and high sugar can harm microbial diversity.

Antibiotics: Kill both harmful and beneficial bacteria. Necessary when prescribed, but overuse can damage the microbiome.

Environmental Chemicals: Some research suggests herbicides like glyphosate may affect gut bacteria because many microbes have the EPSPS enzyme that glyphosate targets. However, real-world evidence in humans is still limited.

Stress: Chronic stress can alter gut bacteria composition.

Sleep: Poor sleep affects the microbiome.

Supporting a Healthy Microbiome:

• Eat diverse plant foods (fruits, vegetables, whole grains, legumes, nuts, seeds)

• Include fermented foods (yogurt, kefir, sauerkraut, kimchi, miso)

• Eat plenty of fiber (aim for 25-35 grams per day)

• Limit ultra-processed foods

• Avoid unnecessary antibiotics

• Manage stress through exercise, sleep, and relaxation

• Spend time outdoors and with animals (exposure to diverse microbes)

Classroom Connection: The gut microbiome is a great example of an ecosystem. Just like a forest or pond, diversity and balance are key to health. Students can understand that what they eat "feeds" their gut microbes, and healthy microbes help them feel good.

Environmental Impacts of Modern Agriculture

Modern industrial agriculture has dramatically increased food production, but it also has significant environmental impacts that are important for students to understand.

Biodiversity Loss:

• Monoculture farming (growing one crop over large areas) reduces plant diversity

• Herbicide use eliminates wildflowers and habitat for pollinators

• Insect populations have declined dramatically in agricultural areas

• Loss of beneficial insects affects pollination and natural pest control

Soil Degradation:

• Intensive tillage destroys soil structure and kills soil organisms

• Loss of organic matter reduces soil's ability to hold water and nutrients

• Erosion washes away topsoil—it takes hundreds of years to form an inch of topsoil

• Compaction from heavy machinery damages soil

Water Pollution:

• Herbicides and pesticides run off into streams, rivers, and groundwater

• Fertilizer runoff causes algae blooms and dead zones in waterways

• Contaminated water affects drinking water supplies and aquatic ecosystems

Climate Change:

• Agriculture contributes about 10% of U.S. greenhouse gas emissions

• Soil degradation releases stored carbon into the atmosphere

• Production of synthetic fertilizers and pesticides requires fossil fuels

• Transportation of food over long distances adds emissions

Regenerative Agriculture: Practices that restore soil health, increase biodiversity, and sequester carbon:

• Cover cropping: Planting crops between growing seasons to protect and enrich soil

• Crop rotation: Changing crops each year to break pest cycles and improve soil

• No-till or reduced-till farming: Minimizing soil disturbance

• Integrated pest management: Using multiple methods instead of relying only on chemicals

• Agroforestry: Integrating trees into farming systems

Organic Farming: Avoids synthetic pesticides and fertilizers, focuses on soil health and biodiversity. Typically has lower yields but better environmental outcomes.

Local and Small-Scale Farming: Often uses more diverse practices, shorter supply chains, and closer relationships between farmers and consumers.

The Challenge: Feeding a growing global population while protecting the environment is one of the great challenges of our time. There are no simple solutions. Students should understand that farmers face economic pressures and alternative methods have tradeoffs, and that systemic changes (policy, subsidies, consumer demand) are needed alongside individual choices.

Section 4: Parent Communication Materials

This section provides ready-to-use materials for communicating with parents about the unit. You can adapt these templates for newsletters, emails, or parent night presentations.

Sample Parent Letter

Date: _____

Dear Families,

Our class is beginning an exciting unit called "From Seed to Spoon: Understanding Our Food System." We will be reading the children's book The Cornfield Maze Mystery and exploring where our food comes from, how it's produced, and how we can make informed choices about what we eat.

This unit is designed to spark curiosity and critical thinking—not to cause alarm. Students will learn about:

• The journey of corn from farm to table
• Modern farming practices, including GMO crops and herbicides
• Food processing and how corn-derived ingredients appear in many products
• Ultra-processed foods and their effects on health
• The gut microbiome and how diet affects our bodies
• Environmental impacts of different farming methods
• Practical actions families can take to support health and sustainability

We will conduct hands-on experiments, including:

• Observing and comparing soil samples
• Testing plant growth in different conditions
• Investigating food labels to identify corn-derived ingredients
• Creating research posters about food choices

This unit aligns with Next Generation Science Standards and Common Core ELA standards. It emphasizes inquiry-based learning, source evaluation, and respectful dialogue about complex issues.

How You Can Support Learning at Home:

• Read the book together and discuss the story
• Involve your child in grocery shopping and meal preparation
• Read food labels together and talk about ingredients
• Visit a farmers market or local farm if possible
• Try one new whole food recipe together
• Ask your child what they're learning and what questions they have

We will be sending home worksheets and activities throughout the unit. Please feel free to reach out if you have any questions or concerns.

Thank you for your support!

Sincerely,

_____

Teacher Name

FAQ for Parents

Q: Is this unit trying to scare kids about food?

A: No. The unit is designed to empower students with knowledge, not to create fear. We present balanced, age-appropriate information and emphasize that people have choices at every level. The focus is on curiosity, critical thinking, and practical actions.

Q: Will my child refuse to eat certain foods after this unit?

A: The unit does not tell students what to eat or avoid. Instead, it teaches them to read labels, understand where food comes from, and make informed choices. We emphasize that occasional consumption of processed foods is fine and that balance is key. If your child expresses concerns about specific foods, use it as an opportunity for family discussion about your values and choices.

Q: Are GMOs safe? What will you teach about them?

A: We teach that GMOs are crops whose DNA has been modified by scientists, that most corn in the U.S. is GMO, and that major scientific organizations consider currently approved GMO crops safe to eat. We also discuss concerns some people have about environmental impacts, corporate control of seeds, and the need for ongoing research. Students learn that this is a topic where people have different perspectives based on their values and interpretation of evidence.

Q: What about herbicides like glyphosate? Are they dangerous?

A: We teach that herbicides are chemicals used to control weeds, that glyphosate is the most common herbicide, and that there is scientific debate about its safety. Regulatory agencies say it's safe at typical exposure levels, while some independent researchers and organizations have concerns. We emphasize that scientists are still studying these questions and that making informed choices means considering multiple sources of evidence. We do not handle any chemicals in class—all activities are observational only.

Q: What are ultra-processed foods?

A: Ultra-processed foods are industrial formulations made mostly from extracted substances (oils, fats, sugars, starches) plus additives. Examples include sodas, chips, instant noodles, and many packaged snacks. Research shows that diets high in ultra-processed foods are associated with obesity, diabetes, heart disease, and other health problems. We teach students to recognize these foods by reading labels and to focus on eating more whole foods (fruits, vegetables, whole grains, beans, nuts) when possible.

Q: Do we need to buy all organic food now?

A: No. Organic food has benefits (fewer synthetic pesticides, better for the environment in some ways), but it's more expensive and not accessible to everyone. We teach that the most important thing is to eat more fruits and vegetables—organic or conventional—and to focus on whole foods rather than ultra-processed foods. Families can make choices based on their budget, values, and priorities.

Q: What if we can't afford to change how we eat?

A: We emphasize that small changes matter and that healthy eating doesn't have to be expensive. Strategies include: buying frozen fruits and vegetables (nutritious and affordable), cooking dried beans and grains (very cheap), reducing soda and packaged snacks (saves money), shopping sales and seasonal produce, and starting with one small swap at a time. The unit is about empowerment and informed choices, not judgment or pressure.

Q: How can I talk to my child about these topics at home?

A: Ask open-ended questions: "What did you learn today?" "What surprised you?" "What questions do you have?" Involve your child in food-related activities like grocery shopping, reading labels, and cooking. Share your family's values and choices without dismissing what they're learning. Model curiosity and critical thinking. If your child asks difficult questions you don't know how to answer, it's okay to say "That's a great question—let's find out together."

Q: Where can I learn more?

A: The companion guide includes a bibliography and list of resources. You can also consult your local Cooperative Extension office, talk to your family doctor or a registered dietitian, and explore reputable sources like the National Academies of Sciences, peer-reviewed journals, and university research programs.

At-Home Extension Activities

These activities help families extend learning beyond the classroom. They are designed to be fun, accessible, and educational.

Activity 1: Pantry Investigation

What you'll need: Food packages from your pantry, paper and pencil

What to do:

1. Choose 10 packaged foods from your pantry

2. Read the ingredient lists together

3. Look for corn-derived ingredients (corn syrup, corn starch, corn oil, dextrose, maltodextrin, etc.)

4. Count how many products contain corn ingredients

5. Discuss: Were you surprised? Why is corn in so many foods?

Activity 2: Whole Food vs. Ultra-Processed Challenge

What you'll need: Your regular groceries

What to do:

1. For one week, try replacing one ultra-processed snack per day with a whole food option

2. Examples: Apple slices instead of chips, carrots and hummus instead of crackers, nuts instead of candy

3. Keep a journal: How do you feel? Do you notice any differences in energy, mood, or hunger?

4. Discuss: Was it hard? What did you like? What would you keep doing?

Activity 3: Start a Small Garden

What you'll need: Seeds (beans, lettuce, herbs), soil, pots or a small garden bed, water

What to do:

1. Choose 2-3 easy-to-grow plants

2. Plant seeds together, following packet directions

3. Care for plants daily (watering, observing)

4. Keep a garden journal with drawings and observations

5. Harvest and eat what you grow!

6. Discuss: What did you learn about where food comes from? What was challenging?

Activity 4: Visit a Farmers Market or Farm

What you'll need: Transportation to a local farmers market or farm

What to do:

1. Visit a farmers market or farm stand

2. Talk to farmers: What do they grow? How do they grow it? What challenges do they face?

3. Buy something you've never tried before

4. Prepare and eat it together

5. Discuss: How is this different from grocery store shopping? What did you learn?

Activity 5: Make a Meal Together

What you'll need: Ingredients for a simple meal using whole foods

What to do:

1. Choose a recipe together (examples: stir-fry with vegetables and rice, pasta with homemade tomato sauce, beef tacos)

2. Shop for ingredients together, reading labels

3. Cook together, with age-appropriate tasks for your child

4. Eat together and discuss: How does this taste compared to packaged versions? Was it fun to make?

5. Make it a weekly tradition!

Activity 6: Compost at Home

What you'll need: Small compost bin or designated outdoor area, food scraps, yard waste

What to do:

1. Set up a simple compost system (many tutorials available online)

2. Collect vegetable scraps, fruit peels, coffee grounds, eggshells, yard waste

3. Avoid meat, dairy, and oils

4. Turn or mix compost weekly

5. Observe changes over weeks and months

6. Use finished compost in your garden

7. Discuss: How does composting reduce waste? How does it help soil?

Activity 7: Food Diary and Reflection

What you'll need: Notebook and pencil

What to do:

1. For one week, keep a simple food diary: What did you eat? How did you feel afterward?

2. Note: Energy level, mood, hunger, stomach comfort

3. At the end of the week, look for patterns

4. Discuss: Do certain foods make you feel better or worse? What changes might you want to make?

5. Remember: This is about awareness, not judgment!

Activity 8: Research a Food Topic Together

What you'll need: Internet access or library

What to do:

1. Choose a topic your child is curious about (examples: Where does chocolate come from? How is cheese made? What are probiotics?)

2. Research together using reputable sources

3. Create a poster, presentation, or report

4. Share what you learned with family or friends

5. Discuss: What surprised you? What questions do you still have?

Section 5: Additional Resources

Expanded Glossary with Pronunciations

This comprehensive glossary includes all key terms from the story and lessons, with pronunciations and age-appropriate definitions. Terms are organized alphabetically.

Additive

Pronunciation: AD-ih-tiv

A substance added to food to preserve it, enhance flavor, or improve appearance. Examples include colors, flavors, and preservatives.

Agriculture

Pronunciation: AG-rih-kul-chur

The science and practice of farming, including growing crops and raising animals for food.

Amino Acid

Pronunciation: uh-MEE-no AS-id

Building blocks of proteins. Living things need amino acids to grow and function.

Atrazine

Pronunciation: AT-ruh-zeen

A herbicide used to control weeds in corn fields. It has been banned in Europe due to water contamination concerns.

Bioaccumulation

Pronunciation: BY-oh-uh-kyoo-myoo-LAY-shun

When a chemical builds up in an organism over time, usually because it's absorbed faster than it can be eliminated.

Biodiversity

Pronunciation: BY-oh-dih-VER-sih-tee

The variety of living things in an ecosystem. More biodiversity usually means a healthier, more resilient ecosystem.

Boom Sprayer

Pronunciation: BOOM SPRAY-er

A machine with long arms that sprays liquids (like herbicides or fertilizers) evenly across a field.

Bt Corn

Pronunciation: BEE-TEE CORN

Corn that has been genetically modified to produce a protein from the bacterium Bacillus thuringiensis. This protein kills certain insect pests.

Carcinogenic

Pronunciation: kar-sin-oh-JEN-ik

Capable of causing cancer.

Combine Harvester

Pronunciation: KOM-bine HAR-ves-ter

A large machine that harvests crops by cutting, threshing, and separating grain from stalks and husks.

Compost

Pronunciation: KOM-post

Decayed organic material (like food scraps and leaves) used to enrich soil and help plants grow.

Contamination

Pronunciation: kon-tam-ih-NAY-shun

The presence of unwanted or harmful substances in something, like chemicals in water or soil.

Conventional Farming

Pronunciation: kon-VEN-shun-al FAR-ming

Modern farming methods that may use synthetic pesticides, herbicides, and fertilizers.

Corn Gluten

Pronunciation: CORN GLOO-ten

The protein part of corn, often used in animal feed or as a food additive.

Corn Starch

Pronunciation: CORN STARCH

A white powder made from corn, used to thicken foods like soups, sauces, and puddings.

Corn Syrup

Pronunciation: CORN SIR-up

A sweet liquid made from corn starch, used as a sweetener in many processed foods.

Cover Crop

Pronunciation: KUV-er KROP

A crop planted between growing seasons to protect soil, prevent erosion, add nutrients, and suppress weeds.

Crop Rotation

Pronunciation: KROP roh-TAY-shun

The practice of growing different crops in the same field in different years to improve soil health and break pest cycles.

Desiccant

Pronunciation: DES-ih-kant

A substance used to dry something out. Some herbicides are used as desiccants to dry crops before harvest.

Dextrose

Pronunciation: DEK-strohs

A simple sugar made from corn, chemically identical to glucose.

Ecosystem

Pronunciation: EE-koh-sis-tem

A community of living things (plants, animals, microorganisms) interacting with each other and their environment.

Endocrine Disruptor

Pronunciation: EN-doh-krin dis-RUP-tor

A chemical that interferes with the body's hormone system, potentially affecting growth, development, and reproduction.

Enzyme

Pronunciation: EN-zime

A protein that speeds up chemical reactions in living things. Enzymes are essential for digestion, growth, and many other processes.

EPSPS

Pronunciation: E-P-S-P-S

An enzyme that plants use to make certain amino acids. Glyphosate works by blocking this enzyme.

Erosion

Pronunciation: ih-ROH-zhun

The wearing away of soil by wind, water, or other natural forces.

Fermented Food

Pronunciation: fer-MEN-ted FOOD

Food that has been transformed by beneficial bacteria or yeast. Examples include yogurt, sauerkraut, and kimchi.

Fertilizer

Pronunciation: FER-tih-ly-zer

A substance added to soil to provide nutrients that help plants grow.

Genetic Engineering

Pronunciation: jeh-NET-ik en-jih-NEER-ing

The process of changing an organism's DNA to give it new traits.

Genetically Modified Organism (GMO)

Pronunciation: jeh-NET-ik-lee MOD-ih-fide OR-gan-iz-um

A plant, animal, or microorganism whose DNA has been altered using genetic engineering.

Glyphosate

Pronunciation: GLY-foh-sate

The most widely used herbicide in the world. Brand name: Roundup. It kills plants by blocking an enzyme they need to grow.

Grain Cart

Pronunciation: GRAYN KART

A large wagon that collects grain from a combine harvester and transfers it to trucks.

Gut Microbiome

Pronunciation: GUT MY-kroh-by-ohm

The community of trillions of bacteria, fungi, and other microorganisms living in the digestive system.

Half-Life

Pronunciation: HAF-life

The time it takes for half of a chemical to break down in the environment.

Herbicide

Pronunciation: HER-bih-side

A chemical used to kill unwanted plants (weeds).

Herbicide-Tolerant

Pronunciation: HER-bih-side TOL-er-ant

Describes a plant that has been genetically modified to survive exposure to a specific herbicide.

High-Fructose Corn Syrup (HFCS)

Pronunciation: HY FRUK-tohs CORN SIR-up

A sweetener made from corn syrup, commonly used in sodas, baked goods, and many processed foods.

Hyperpalatable

Pronunciation: HY-per-PAL-uh-tuh-bul

Extremely tasty and appealing, often engineered to encourage overconsumption.

Inflammation

Pronunciation: in-fluh-MAY-shun

The body's response to injury or infection, involving redness, swelling, and sometimes pain. Chronic inflammation can contribute to disease.

Insecticide

Pronunciation: in-SEK-tih-side

A chemical used to kill insects that damage crops.

Integrated Pest Management (IPM)

Pronunciation: IN-teh-gray-ted PEST MAN-ij-ment

An approach to pest control that uses multiple methods (beneficial insects, crop rotation, targeted spraying) instead of relying only on chemicals.

Litigation

Pronunciation: lit-ih-GAY-shun

The process of taking legal action; lawsuits.

Maltodextrin

Pronunciation: mal-toh-DEK-strin

A powder made from corn, rice, or potato starch, used as a filler or thickener in processed foods.

Maximum Residue Limit (MRL)

Pronunciation: MAX-ih-mum REZ-ih-doo LIM-it

The highest amount of a pesticide legally allowed to remain in or on food, set by regulatory agencies.

Metabolite

Pronunciation: meh-TAB-oh-lite

A substance formed when a chemical breaks down in the environment or in a living organism.

Microbe

Pronunciation: MY-krohb

A tiny living thing, such as a bacterium or fungus, that can only be seen with a microscope.

Microbiome

Pronunciation: MY-kroh-by-ohm

The community of microorganisms living in a particular environment, such as soil or the human gut.

Modified Corn Starch

Pronunciation: MOD-ih-fide CORN STARCH

Corn starch that has been chemically or physically altered to change its properties, used in many processed foods.

Monoculture

Pronunciation: MON-oh-kul-chur

The practice of growing a single crop over a large area, year after year.

Organic Farming

Pronunciation: or-GAN-ik FAR-ming

Farming without synthetic pesticides, herbicides, or fertilizers. Organic farmers use natural methods to build soil health and control pests.

Organic Matter

Pronunciation: or-GAN-ik MAT-er

Material that comes from living things, such as decayed plants and animals. Organic matter enriches soil.

Pesticide

Pronunciation: PES-tih-side

A chemical used to kill pests, including insects, weeds, fungi, and rodents.

Planter

Pronunciation: PLAN-ter

A machine that plants seeds in neat rows at precise depths and spacing.

Pollinator

Pronunciation: POL-ih-nay-tor

An animal (like a bee, butterfly, or bird) that carries pollen from one flower to another, helping plants reproduce.

Processing Plant

Pronunciation: PROH-ses-ing PLANT

A factory where raw crops are turned into ingredients or products used in food manufacturing.

Regenerative Agriculture

Pronunciation: rih-JEN-er-uh-tiv AG-rih-kul-chur

Farming practices that restore soil health, increase biodiversity, and capture carbon from the atmosphere.

Residue

Pronunciation: REZ-ih-doo

A small amount of a substance that remains after most of it has been removed or used up.

Shikimate Pathway

Pronunciation: SHIK-ih-mate PATH-way

A series of chemical reactions that plants and some microorganisms use to make certain amino acids. Glyphosate blocks this pathway.

Soil Organic Matter

Pronunciation: SOYL or-GAN-ik MAT-er

Decayed plant and animal material in soil that feeds soil organisms and helps soil hold water and nutrients.

Subsidy

Pronunciation: SUB-sih-dee

Money given by the government to support farmers or industries, often affecting which crops are grown and how much they cost.

Synergistic Effect

Pronunciation: sin-er-JIS-tik ih-FEKT

When two or more substances combine to produce an effect greater than the sum of their individual effects.

Tillage

Pronunciation: TIL-ij

The practice of turning over soil to prepare it for planting. Excessive tillage can damage soil structure.

Ultra-Processed Food

Pronunciation: UL-truh PROH-sest FOOD

Food made mostly from industrial ingredients (oils, sugars, starches, proteins) with many additives. Examples: sodas, chips, instant noodles.

Water-Soluble

Pronunciation: WAH-ter SOL-yoo-bul

Able to dissolve in water. Water-soluble chemicals can wash away easily but may also contaminate water supplies.

Weed

Pronunciation: WEED

A plant growing where it's not wanted, competing with crops for water, nutrients, and sunlight.

Wet Milling

Pronunciation: WET MIL-ing

A process that soaks corn kernels in water to soften them, then separates them into parts (germ, fiber, gluten, starch) for different uses.

Bibliography and Further Reading

For Educators:

• National Academies of Sciences, Engineering, and Medicine. (2016). Genetically Engineered Crops: Experiences and Prospects. Washington, DC: The National Academies Press.

• International Agency for Research on Cancer (IARC). (2015). IARC Monographs Volume 112: Evaluation of five organophosphate insecticides and herbicides.

• Monteiro, C.A., et al. (2018). Ultra-processed foods: what they are and how to identify them. Public Health Nutrition, 22(5), 936-941.

• Mesnage, R., & Antoniou, M.N. (2020). Computational modelling provides insight into the effects of glyphosate on the shikimate pathway in the human gut microbiome. Current Research in Toxicology, 1, 25-33.

• Valdes, A.M., et al. (2018). Role of the gut microbiota in nutrition and health. BMJ, 361, k2179.

• Your local Cooperative Extension office (search '[your state] cooperative extension' for resources, soil testing, and local farming information)

For Students and Families:

• Pollan, Michael. (2009). The Omnivore's Dilemma for Kids: The Secrets Behind What You Eat. Dial Books.

• Bang, Molly, & Chisholm, Penny. (2009). Living Sunlight: How Plants Bring the Earth to Life. Blue Sky Press.

• Ehlert, Lois. (1987). Growing Vegetable Soup. Harcourt Brace Jovanovich.

• Gibbons, Gail. (2008). The Vegetables We Eat. Holiday House.

• Peterson, Cris. (2010). Seed Soil Sun: Earth's Recipe for Food. Boyds Mills Press.

- USDA National Agricultural Library: www.nal.usda.gov

- National Farm to School Network: www.farmtoschool.org

- KidsGardening: www.kidsgardening.org

- Environmental Working Group (EWG): www.ewg.org

- Center for Science in the Public Interest: www.cspinet.org

- Local Harvest (find farmers markets and CSAs): www.localharvest.org

Community Connections

Connecting students with local food systems and experts enriches learning and builds community relationships. Here are suggestions for bringing the food system to life:

Guest Speakers:
- Local farmers (conventional and organic)

- Cooperative Extension agents

- Registered dietitians or nutritionists

- Food safety inspectors

- Grocery store managers

- Community garden coordinators

- Soil scientists

Field Trips:
- Local farms (arrange tours during planting or harvest)

- Farmers markets

- Food processing facilities

- Community gardens

- Grocery stores (behind-the-scenes tours)

- Composting facilities

- University agricultural research stations

Service Learning Projects:

• Start a school garden

• Organize a food drive focusing on whole foods

• Create educational materials for younger students

• Host a community food and farming fair

• Partner with a local food bank to sort and distribute produce

• Establish a school composting program

Cooperative Extension Office:

Your local Cooperative Extension office is an invaluable resource. Extension offices are part of the land-grant university system and provide research-based information on agriculture, gardening, nutrition, and more. Services often include:

• Free or low-cost soil testing

• Educational programs and workshops

• Master Gardener volunteers who can visit classrooms

• Curriculum materials and lesson plans

• Information about local farming practices and challenges

• 4-H youth development programs

To find your local office: Search '[your state] cooperative extension' or visit www.nifa.usda.gov/land-grant-colleges-and-universities-partner-website-directory

Extension Projects

These projects allow students to dive deeper into topics of interest and apply their learning in creative ways.

Project 1: Design a Sustainable School Lunch Program

Students research and design a school lunch program that prioritizes whole foods, local sourcing, and minimal waste. They create menus, calculate costs, identify local suppliers, and present their plan to school administrators.

Project 2: Create a Food System Documentary

Students work in teams to create short video documentaries exploring different aspects of the food system: farming practices, food processing, grocery store operations, or family food choices. They conduct interviews, film footage, and edit their work.

Project 3: Conduct a School-Wide Food Audit

Students survey what foods are available at school (cafeteria, vending machines, fundraisers) and analyze them for nutritional quality, processing level, and corn-derived ingredients. They present findings and recommendations to school leadership.

Project 4: Start a Pollinator Garden

Students research native plants that support pollinators, design a garden, raise funds for materials, plant and maintain the garden, and create educational signage. They track pollinator visits and share data with citizen science projects.

Project 5: Investigate Local Water Quality

Students research local water sources, test water samples for contaminants (with appropriate supervision and equipment), map agricultural areas and potential runoff, and present findings to the community.

Project 6: Create a Community Cookbook

Students collect family recipes that use whole foods and minimal processed ingredients. They compile recipes into a cookbook, including nutritional information, cost estimates, and stories about food traditions. Cookbooks can be sold to raise funds for the school garden or donated to families.

Project 7: Organize a Food and Farming Fair

Students plan and host a community event featuring local farmers, food demonstrations, cooking classes, garden tours, and educational booths. This project integrates research, communication, event planning, and community engagement.

Project 8: Develop a Food Literacy Campaign

Students create posters, social media content, videos, or presentations to teach others about reading food labels, understanding ingredients, and making informed choices. They can target different audiences: younger students, families, or the broader community.

The Cornfield Maze Mystery invites students on a journey of discovery—from cornfields to processing plants, from grocery stores to their own bodies. Along the way, they learn that food systems are complex, that science doesn't always provide simple answers, and that everyone has power to make informed choices.

This companion guide provides the tools to facilitate that journey: lesson plans, experiments, discussion prompts, background information, and resources for continued learning. But the most important tool is curiosity—the willingness to ask questions, seek evidence, listen to different perspectives, and think critically about the world.

As educators, we have the privilege of nurturing that curiosity. We can help students see that they are not passive consumers but active participants in the food system. They can read labels, ask questions, grow food, support local farmers, reduce waste, and share what they learn with others.

The Corn Crew—Mary, Ana, and Ben—discovered that asking questions is how we learn, that small actions matter, and that change begins with curiosity. May your students discover the same.

Thank you for using this guide. We hope it sparks meaningful conversations, hands-on learning, and lasting change in your classroom and community.

Download free PDF for worksheets and additional materials:
www.thecornfieldmazemystery.com

Happy teaching, and happy learning!